AEROGELS

and Other Material Tech

Co-published by agreement between Shi Tu Hui and World Book, Inc.

Shi Tu Hui
Room 1807, Block 1,
#3 West Dawang Road
Chaoyang District, Beijing 100025
P.R. China

World Book, Inc
180 North LaSalle Street
Suite 900
Chicago, Illinois 60601
USA

Library of Congress Cataloging-in-Publication Data for this volume has been applied for.

Cool Tech (set #2)
ISBN: 978-0-7166-5387-5 (set, hc)

Aerogels and Other Material Tech
ISBN: 978-0-7166-5388-2 (hc)

Also available as:
ISBN: 978-0-7166-5394-3 (e-book)
ISBN: 978-0-7166-5400-1 (soft cover)

Written by William D. Adams

STAFF

VP, Editorial: Tom Evans

Manager, New Product: Nicholas Kilzer

Curriculum Designer: Caroline Davidson

Proofreader: Nathalie Strassheim

Indexer: Shawn Brennan

Coordinator, Design Development & Production: Brenda Tropinski

Senior Media Editor: Rosalia Bledsoe

Developed with World Book by White-Thomson Publishing LTD

www.wtpub.co.uk

ACKNOWLEDGMENTS

Cover   © H.S. Photos/Alamy Image

5   MSFC/NASA

6-7   NASA/JPL-Caltech

8-9   © Beauty Image/Shutterstock; Utah State History/J. Willard Marriott Library, University of Utah; © Monsanto Chemical Company; © LuYago/Shutterstock

10-11   NASA/JPL-Caltech; NASA; NASA/JPL-Caltech/MSSS; © Elite

12-13   © Triff/Shutterstock; JSC/NASA; Aspen Systems Inc/NASA; © Siwakorn1933/ Shutterstock; © Aspen Aerogels

14-15   © 3d_illustrator/Shutterstock; © BONNINSTUDIO/Shutterstock

16-17   © sumire8/Shutterstock; © Ruslan Ivantsov, Shutterstock; © remedios55/Shutterstock; Nobel Prize Museum

18-19   © Luciano Santandreu, Shutterstock; © dandesign86/Shutterstock; University of Arkansas; © Kaspars Grinvalds, Shutterstock

20-21   © Tupungato/Shutterstock; JSC/NASA; © Zhejiang University

22-23   © nobeastsofierce/Shutterstock; © Mopic/Shutterstock

24-25   © Angel Soler Gollonet, Shutterstock; © Bao Lab/Stanford University; © Emaar; © Greenshoots Communications/Alamy Images

26-27   NASA/Pat Rawlings

28-29   © Lukasz Libuszewski, Shutterstock; © Ashley Cooper, Alamy Images

30-31   © Pavel Ilyukhin, Shutterstock; © Sergey Kamshylin, Shutterstock; © Tea Talk/Shutterstock; Library of Congress; Public Domain

32-33   © Kobets Dmitry, Shutterstock; © Mykhailo Pavlenko, Shutterstock

34-35   © J. P. Oleson, University of Utah; © Madredus/Shutterstock; © Sika; © Bannafarsai Stock/Shutterstock; © Carbicrete; © Arcaid Images/Alamy Images

36-37   © luchschenF/Shutterstock; © REUTERS/Alamy Images

38-39   © Pixinoo/Shutterstock; WORLD BOOK illustration; © E. R. Degginger; © ONYXprj/Shutterstock; World Book illustration by Richard Bonson, The Art Agency

40-41   © Vladimirkarp/Shutterstock; © Cornel Constantin, Shutterstock; Radislav A. Potyrailo et al.; © Italay/Shutterstock

42-43   U.S. Air Force photo/Staff Sgt. Bennie J. Davis III

44-45   © Edgehog Advanced Technologies; © David Schurig, Duke University; © Xiang Zhang, Berkeley Lab; © Sergey Nivens, Shutterstock; © Parabol Studio/Shutterstock; © Junfei Li; © Jiangfeng Zhou and Clayton Fowler

CONTENTS

There is a glossary of terms on the first page. Terms defined in the glossary are in boldface type that **looks like this** on their first appearance on any spread (two facing pages).

GLOSSARY

aerogel an artificial, dry solid known for its exceptionally low density.

aerographene an aerogel formed from graphene.

architect a person who designs and lays out plans for buildings.

asteroid a rocky or metallic object smaller than a planet that orbits a star.

atmosphere the mass of gases that surrounds a planet or other heavenly body.

carbon nanotube a nanoscale tube made up of atoms of carbon.

engineer a professional who plans and builds engines, machines, roads, or the like.

graphene a sheet of carbon atoms in the form of a hexagonal lattice one atom thick. Graphene is the substance that makes up graphite.

graphite a soft, black form of carbon with a metallic luster.

polyester any one of a large group of synthetic materials made from long, chainlike molecules. Polyesters are used in the manufacture of paints, synthetic fibers, films, and reinforced plastics for construction.

silica a common mineral, silicon dioxide, a hard, white, or colorless substance.

synthetic made artificially.

3D printing a manufacturing technology that creates three-dimensional (3D) objects from computer instructions.

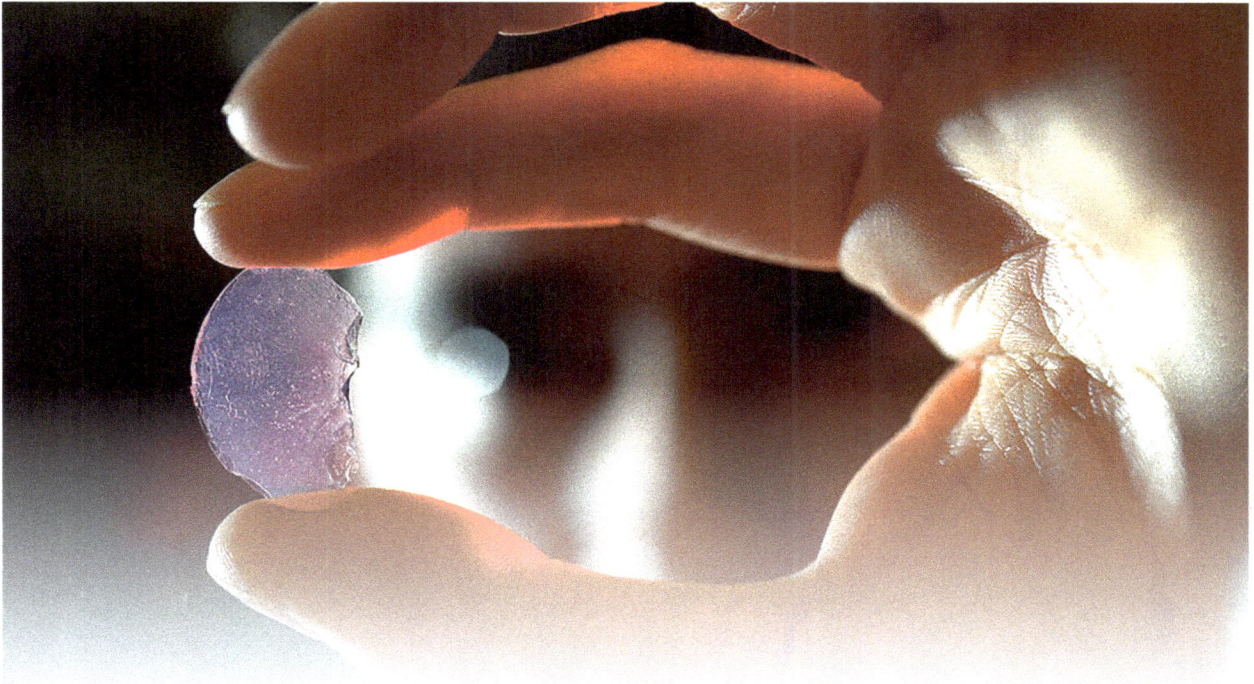

INTRODUCTION

The products that we make and use help to shape our lives and the world around us. But what we can make—and how well we can make it—depends on the materials available.

Early humans used materials they found in nature, such as stone, wood, plant fibers, bone, and animal pelts. Our ancestors used these seemingly simple materials to make countless useful products. But over time, they found ways to improve these materials and create new ones. Learning to extract and work metals revolutionized every facet of human life. Further discoveries of new materials have sparked waves of changes in the products we make and how we make them.

In our modern world of plastics, metal alloys, and semiconductors, it may seem as if we have all the materials we could need. But materials scientists continue to discover new materials with stunning properties, surpassing existing materials in almost every respect. What kind of products will inventors dream up that use these materials to their fullest potential?

1 AEROGELS

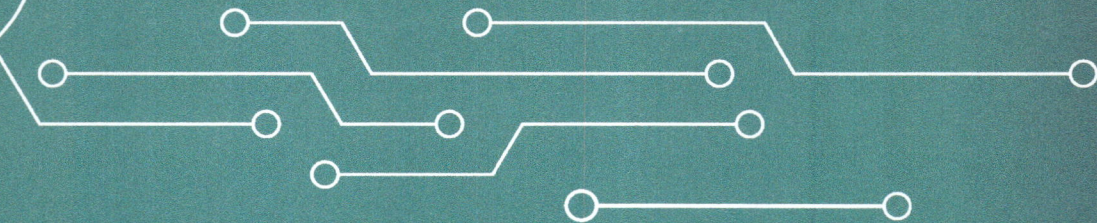

FROZEN SMOKE

Imagine a material so light that it has been described as "frozen smoke." It looks like a block of frosted glass. But it weighs little more than the air around it. This material is an **aerogel,** an artificially made, dry solid with exceptionally low density.

Like a sponge, an aerogel contains many tiny holes or *pores*. These pores measure just *nanometers* (billionths of a meter) in size. There are so many of them, however, that they make up most of an aerogel's bulk. Aerogels typically consist of from 95 percent to more than 99 percent air.

Aerogels are more than a parlor trick of materials science. They have already been used in some of the most challenging environments in the solar system—including the surface of Mars and the depths of Earth's oceans. Soon, they could be coming to your home.

THE DEVELOPMENT OF AEROGELS

According to legend, aerogels were invented as the result of a bet. The American materials scientists Steven Kistler and Charles Learned competed to see who could remove the liquid from a jar of jelly without causing the jelly to collapse. Kistler won the bet. His findings were published in the scientific journal *Nature* in 1931.

The product of gel. A gel is a mixture of liquid and solid. The solid elements form an open network of pores that holds the liquid in place. Gels have a lot of uses—from medicine delivery to some batteries to fruit spreads—but they can be messy and are relatively heavy.

How aerogels are made. If the liquid in a gel simply evaporates, the solid component shrinks and collapses. Aerogels therefore require chemical sleight-of-hand to produce. The gel is placed in a special chamber called an *autoclave.* The autoclave raises it to a high temperature and pressure. The liquid becomes a *supercritical fluid,* a state of matter with many of the properties of a gas as well as a liquid. When the pressure is lowered in the autoclave, the supercritical fluid becomes a gas and is vented away.

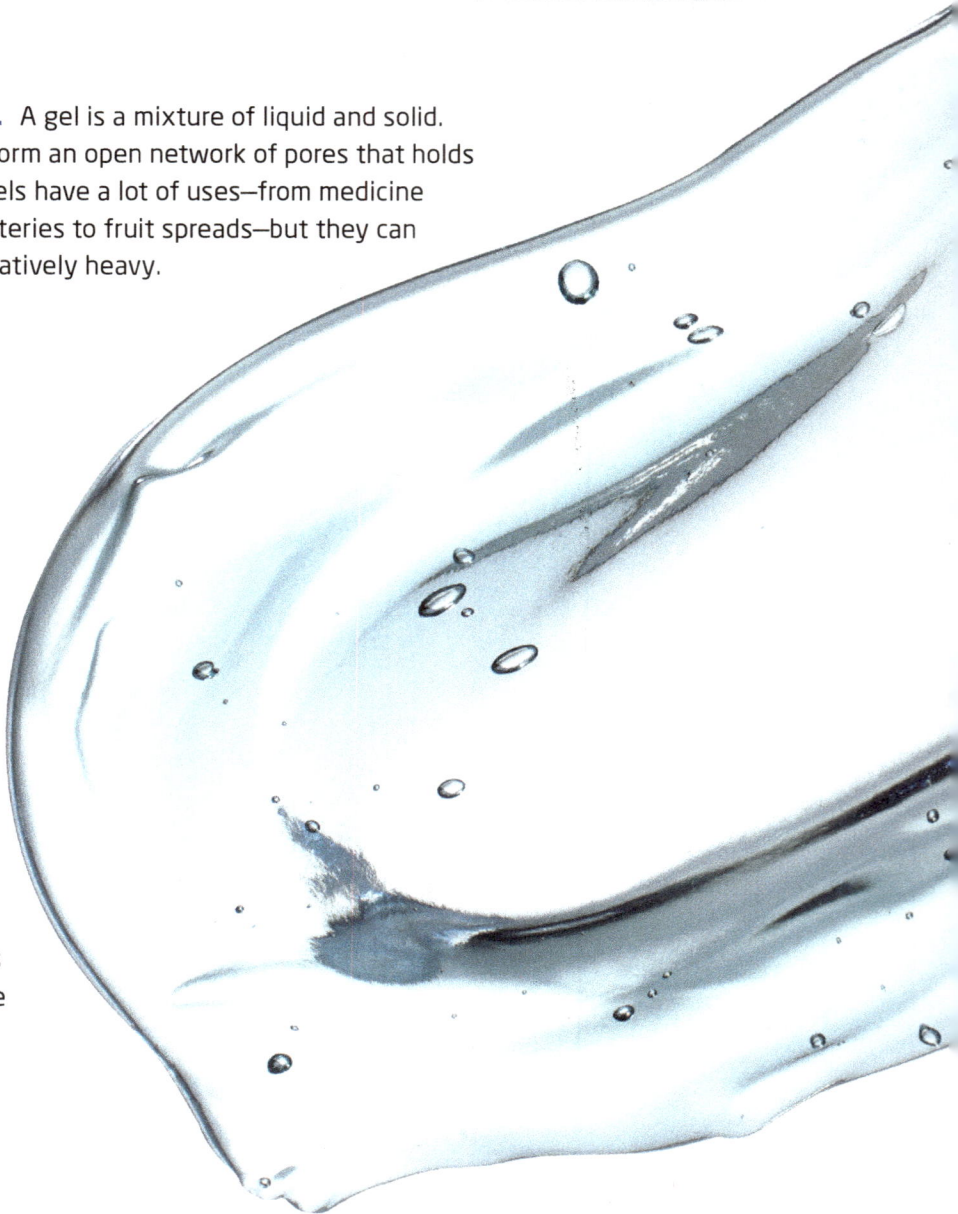

The first products. After Kistler invented aerogels, he went to work at the chemical company Monsanto. There, he led the development of aerogel additives for such products as paint and toothpaste. But other materials were developed for these applications, and Monsanto stopped manufacturing aerogels in the 1970's.

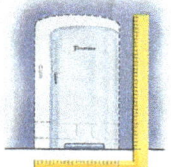
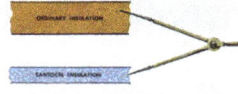

Breakthrough. Monsanto abandoned aerogels in part because they were expensive, time-consuming, and dangerous to produce. But in the late 1970's, materials scientists at the Claude Bernard University in Lyons, France, developed a gel base that made manufacturing much easier.

MODERN AEROGELS

Aerogels aren't just lightweight—they also make amazing insulators. Heat spreads directly through a solid material by conduction. But, aerogels are mostly air, requiring heat to pass by *convection,* the movement of heated air. The solid matrix provides a complex, winding path—like a maze—making it difficult for heated air to get far. This arrangement greatly limits the transfer of heat through the aerogel.

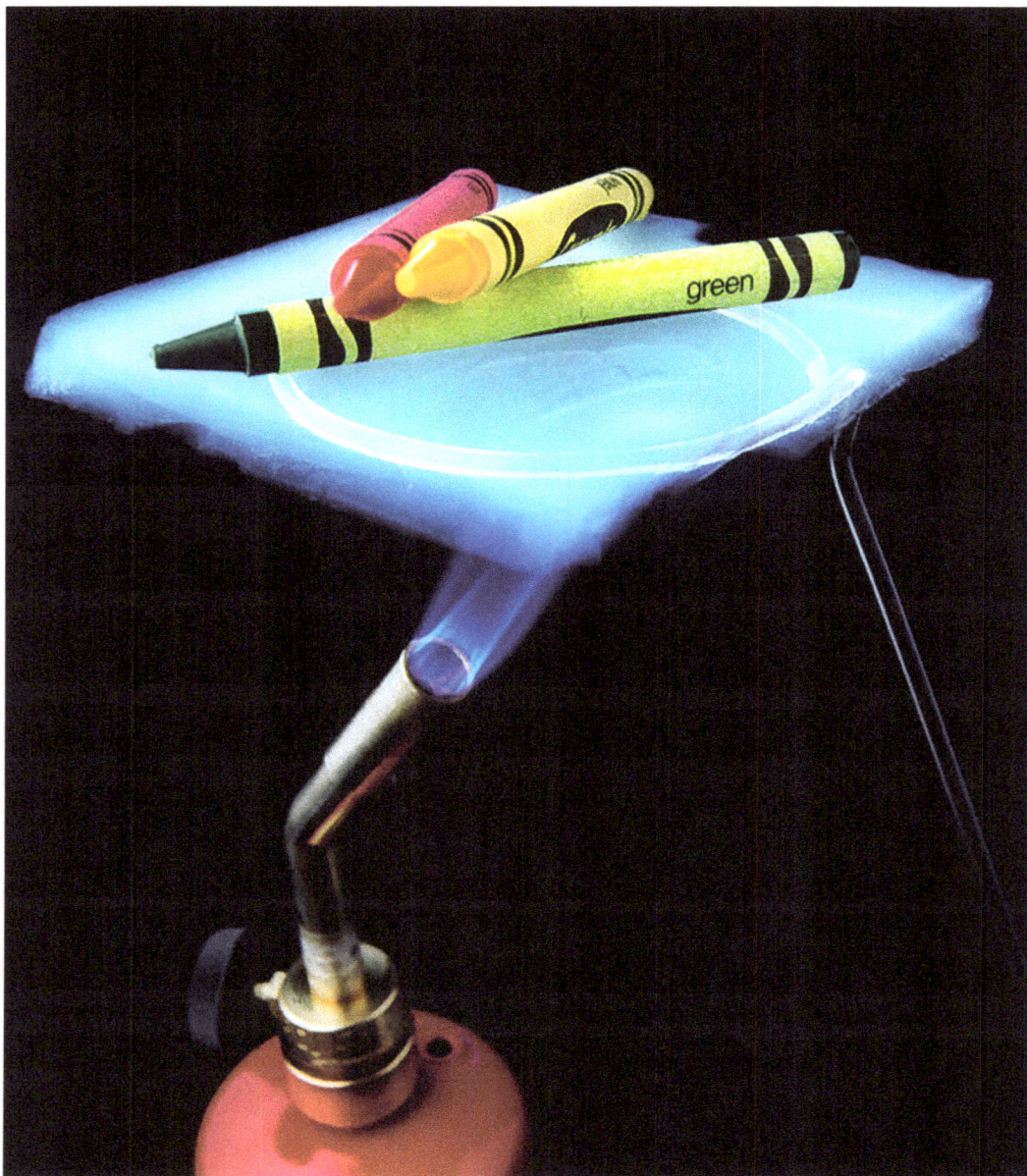

Cosmic dust collector. Studying cosmic dust helps astronomers better understand the origins of the solar system. But cosmic dust is difficult to capture because it travels so fast, breaking down or disintegrating when it collides with collector material. In 1999, the United States launched the Stardust probe with an aerogel dust collector. The porous aerogel crushed on impact from dust grains, gently slowing them down.

Hi-tech insulator on Earth. The use of aerogel is not confined to the plains of Mars. Designers have incorporated aerogel into expensive, high-quality jackets and water bottles. Aerogel jackets provide extreme warmth in a thin jacket. Aerogel-insulated bottles keep water or sports drinks cold while still maintaining their squeezability.

Mars rover insulator. Bitterly cold temperatures on the planet Mars can freeze electronic and scientific equipment, creating the need to insulate spacecraft. But, space and weight are tightly limited in space launches, and traditional insulation is bulky and relatively heavy. For this reason, every U.S. Mars rover since the Mars Pathfinder mission, launched in 1996, has made use of slender, lightweight aerogel insulation.

THE SKY'S THE LIMIT

The mineral **silica** is the most commonly used solid component in aerogels. It is easy to work with and extremely cheap. On its own, silica produces a brittle, glasslike aerogel that isn't right for all applications. But aerogels aren't limited to silica. Kistler reported making aerogels out of many materials—including egg whites! Materials scientists today are crafting aerogels out of an ever-increasing variety of materials.

Flexibility. Silica aerogel is inflexible. Manufacturers may get around this limitation by grinding up aerogel and embedding it in another material. But this procedure adds manufacturing difficulty and reduces insulation. An American company called Aspen Aerogels produces flexible aerogel products through the addition of **polyester** fibers. Their products are used to insulate everything from buildings to spacesuits.

Clean and green. Researchers at the University of Singapore have created an aerogel out of paper waste. This cellulose-based aerogel could be used in environmentally friendly packaging and insulation. It could also be used to clean up oil spills—the porous structure of aerogels makes them good at absorbing oil and other liquids. Because it is made of cellulose, the aerogel would be nontoxic if some of it escaped into the environment.

Under the sea. Aerogel has also been used to insulate oil and natural gas pipelines on the sea floor. Oil and gas pipes must be insulated to keep the icy chill of the water from slowing the flow of the fluid inside. Efficient, lightweight aerogel insulation may only need to be half the thickness of other insulators. The pipes can thus be smaller in diameter, enabling smaller ships to lay them.

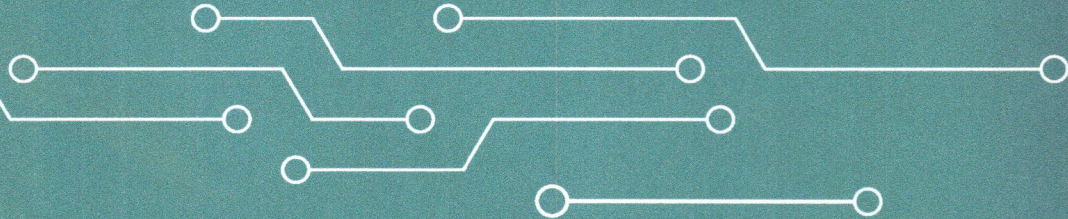

2 GRAPHENE

SUPER-THIN SHEETS OF CARBON

Carbon is a versatile chemical element. It is the main component of coal, diamonds, pencil lead, and even you!

The key to carbon's variety of forms lies in the atom's ability to maintain up to four chemical bonds. This capability enables carbon atoms to be arranged in grids, rings, and long chains—among other structures—and still have plenty of bonds left to include other elements. The result is a stunning diversity of compounds and materials.

Few of these materials are more amazing than **graphene.** Graphene consists of a single layer of carbon atoms arranged in a honeycomb structure. Working with this infinitesimally thin fabric is extremely difficult. But, the payoff could revolutionize nearly every aspect of modern technology.

A CRASH COURSE ON CARBON

Carbon can link up with other carbon atoms and hydrogen, oxygen, and nitrogen atoms to form long chain molecules. Such chains include many of the organic molecules that make up living things. But carbon can also bond exclusively with other carbon atoms, forming diamond and graphite. In diamond, a carbon atom bonds with atoms all around it, forming one of the hardest substances known.

In graphite, carbon atoms form a series of flat, honeycomb-patterned sheets. The sheets are stacked loosely on top of one another, giving graphite (used in pencil lead) its soft, slippery texture. Graphene is a single, isolated layer of graphite.

How to make graphene. Believe it or not, graphene can be made with household materials. One way is to take clear sticky tape and apply it to a block of graphite. Peel the tape off, then apply fresh tape to the original tape. Repeat this process several times. The tape peels the layers of graphite apart until only a single layer remains. The final piece of tape can then be dissolved away with a special detergent. The Russian-born physicists Andre Geim and Konstantin Novoselov won the 2010 Nobel Prize for their discovery of this method.

Out of sticky tape? Use a blender instead! Another team of researchers mixed graphite powder, dishwashing detergent, and water in a blender. The result was a large number of graphene flakes in the water. (Don't try this at home—you'll ruin the blender.) This method of making graphene is relatively easy, giving manufacturers hope of one day making graphene in large quantities.

USES OF GRAPHENE

Water filtration and desalination. One out of three people worldwide still lacks access to clean drinking water. Graphene's honeycomb structure is large enough to allow water molecules through. But, graphene's pores are too small to admit other particles, including heavy metals and salt. Graphene thus presents a tantalizing option for filtering and *desalinating* (removing the salt from) water.

Flexible electronics. Graphene components could help make electronics more flexible. A tablet or smartphone made of graphene electronics printed behind a plastic screen could be as thin and flexible as a sheet of cardboard.

Transistors. Smartphones, computers, and most other electronics rely on *transistors*—tiny devices that control the flow of electric current. Today, almost all transistors are made from silicon. Silicon is cheap, but it is inflexible and can only be made so thin. **Engineers** are working to develop graphene-based transistors for widespread use. Such transistors would assist in making flexible, wearable electronics and help other devices to shrink even further.

Low-voltage energy generators. Graphene practically creates low-voltage electricity out of thin air. In 2020, a team from the University of Arkansas found it possible to create special graphene circuits to harvest energy from small movements in its atoms—called Brownian motion. Many scientists had previously thought it impossible to harvest energy from Brownian motion.

AEROGRAPHENE

Remember that aerogels can be made of almost any material. What if an aerogel was made of graphene? The resulting material is called **aerographene.** Aerographene has some interesting properties that could be used in air purification.

Graphene can be electrically conductive. Running electricity through graphene causes it to heat up, much like the filament in an incandescent light bulb.

When electricity is passed through aerographene, the graphene solid quickly heats to high temperatures. Because the graphene has so much surface contact with the air, the air inside heats quickly as well. The heated air expands and puffs out of the aerographene. When electric current is removed, the aerographene cools as quickly as it heated, allowing room-temperature air to flow back in. By switching electric current on and off, engineers can thus create an air pump with no moving parts.

The high heat inside the aerographene kills viruses, molds, and bacteria and destroys allergens. Simple aerographene air purifiers could be installed in hospitals, airports, and other public spaces to reduce the spread of infectious diseases.

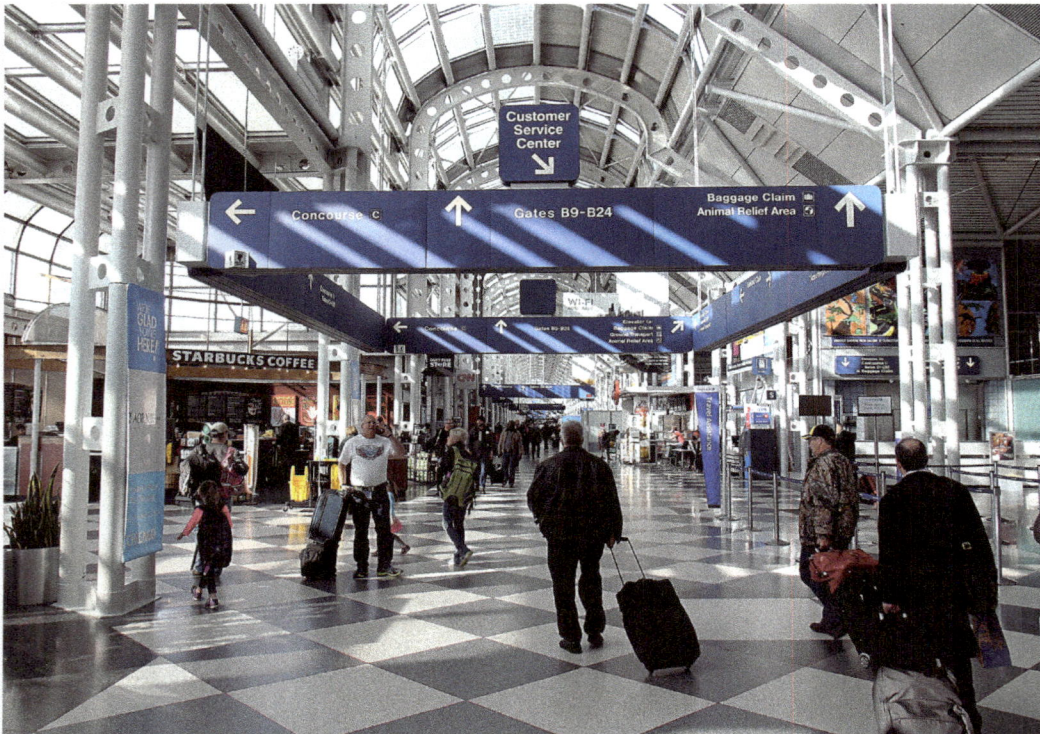

A scanning electron microscope (SEM) image of an aerographene, right, shows a spongelike texture filled pores. Below, a chunk of aerographene is light enough to balance on a flower's petals. The material has an extremely low density, just 1/10 ounce per cubic inch (0.2 grams per cubic centimeter).

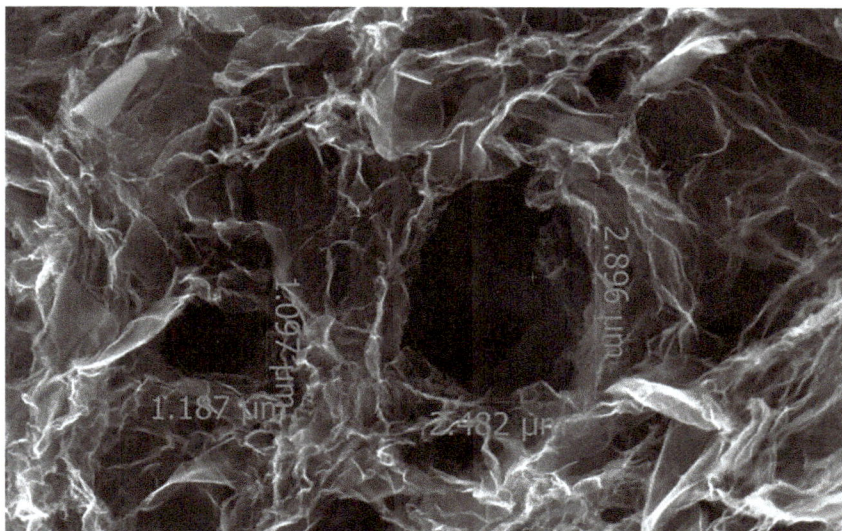

3 CARBON NANOTUBES

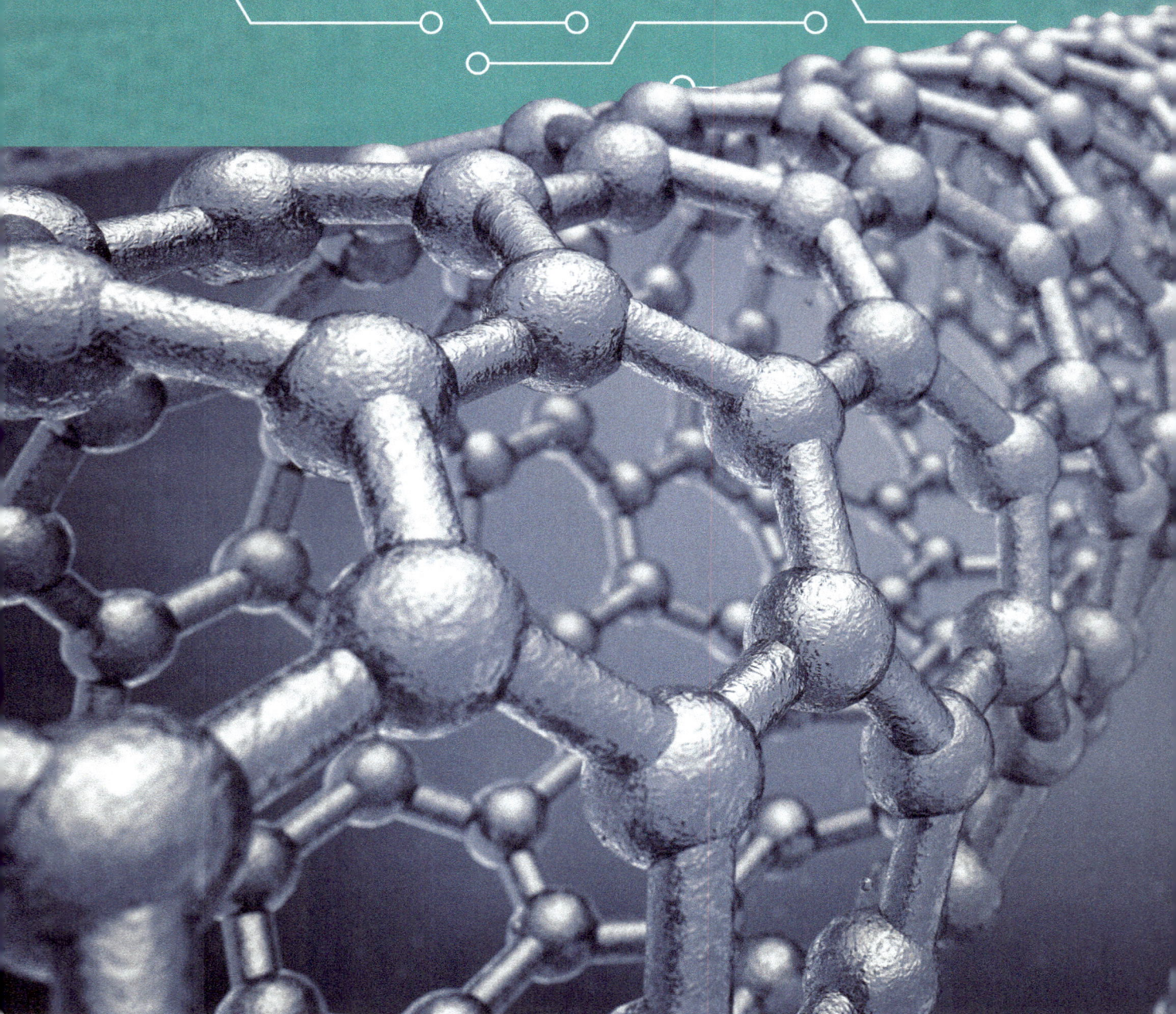

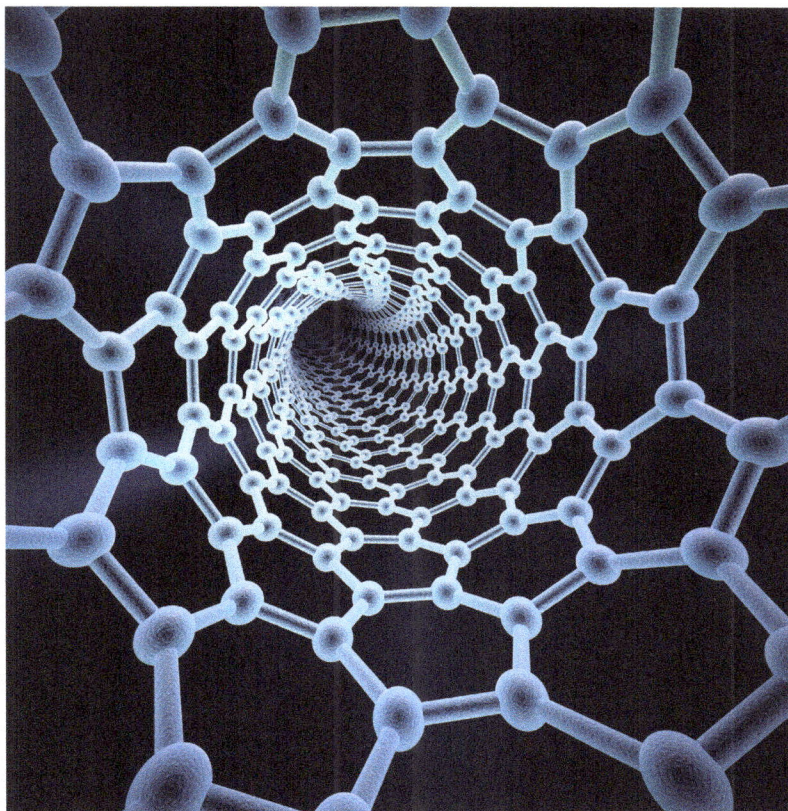

NEVER-ENDING GRAPHENE

Graphene is an amazing material. But every sheet of graphene must end. What if a sheet of graphene could be wrapped around and connected to itself? If you could do it—and it can be done—you would have a **carbon nanotube.**

Nanotubes are tubular structures of carbon atoms several nanometers in diameter and several thousand nanometers in length. Just like a rolled-up sheet of graphene, each nanotube is a single molecule of carbon, with all the atoms bound together in a honeycomblike structure. Nanotubes are many times stronger than steel.

Carbon nanotubes share many properties with graphene, but they also have amazing qualities of their own. And just like graphene, nanotubes are poised to revolutionize materials science.

PROPERTIES AND USES OF NANOTUBES

A carbon nanotube encloses a tiny area inside it. And, carbon nanotubes can vary in diameter. This means that several carbon nanotubes can be nested one within another. The result is a *multiwalled* carbon nanotube. A carbon nanotube can also hold medium-sized molecules too large to escape through the carbon lattice. Each end of the nanotube is capped with a dome-shaped lattice of carbon atoms.

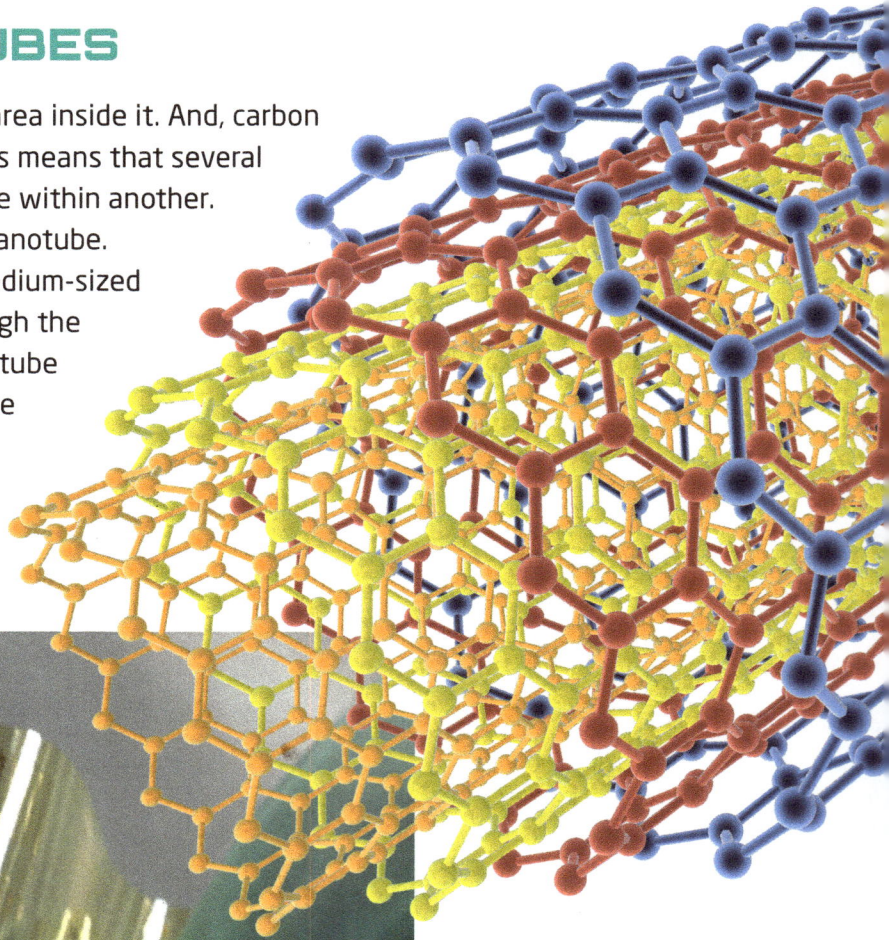

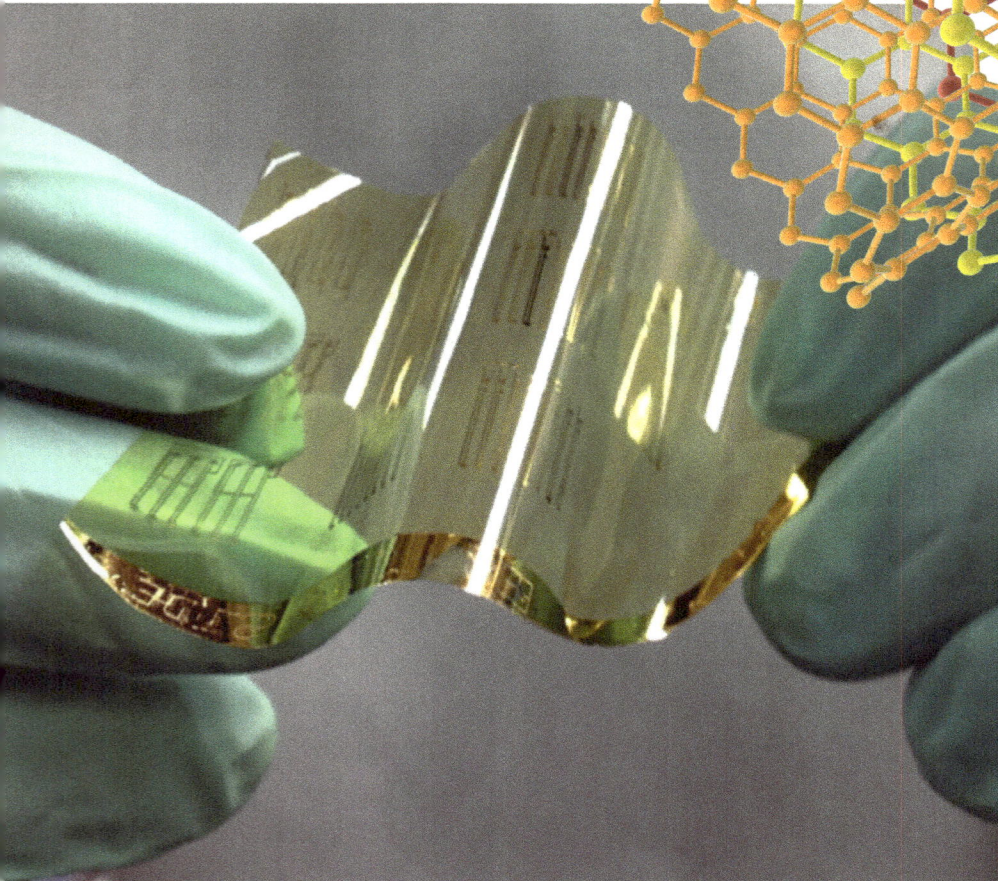

Conductivity. Like graphene, carbon nanotubes can conduct electricity. They could thus be used to create future generations of computer chips. Because they are flexible, nanotubes could even be woven into clothing to charge personal electronic devices on the go.

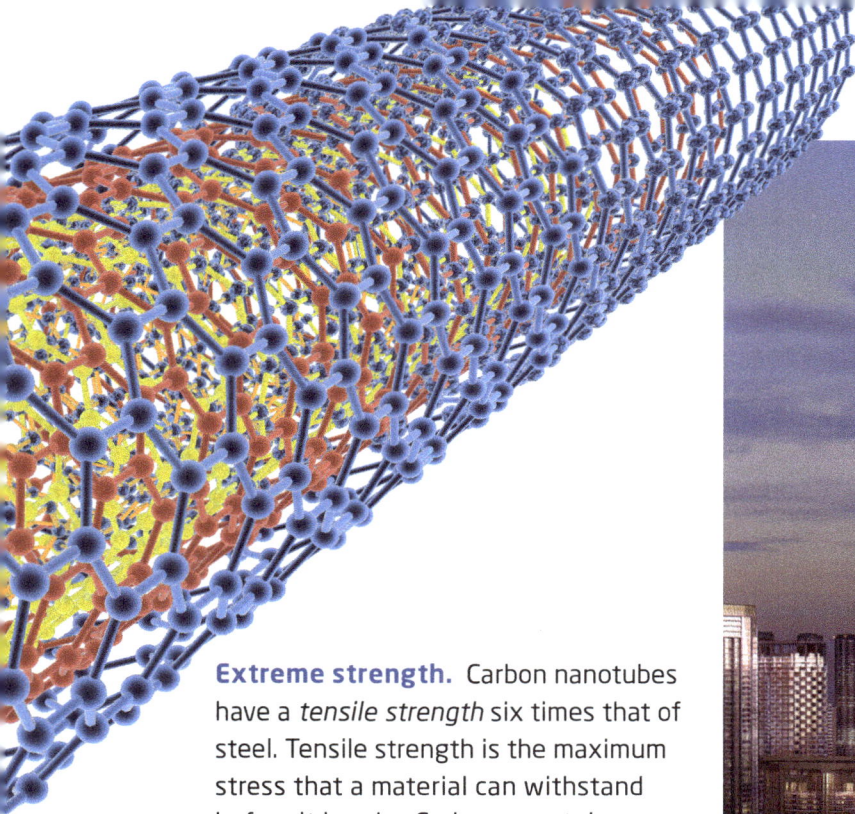

Extreme strength. Carbon nanotubes have a *tensile strength* six times that of steel. Tensile strength is the maximum stress that a material can withstand before it breaks. Carbon nanotubes could enable engineers to create structures that are colossal and strong yet gossamer-thin.

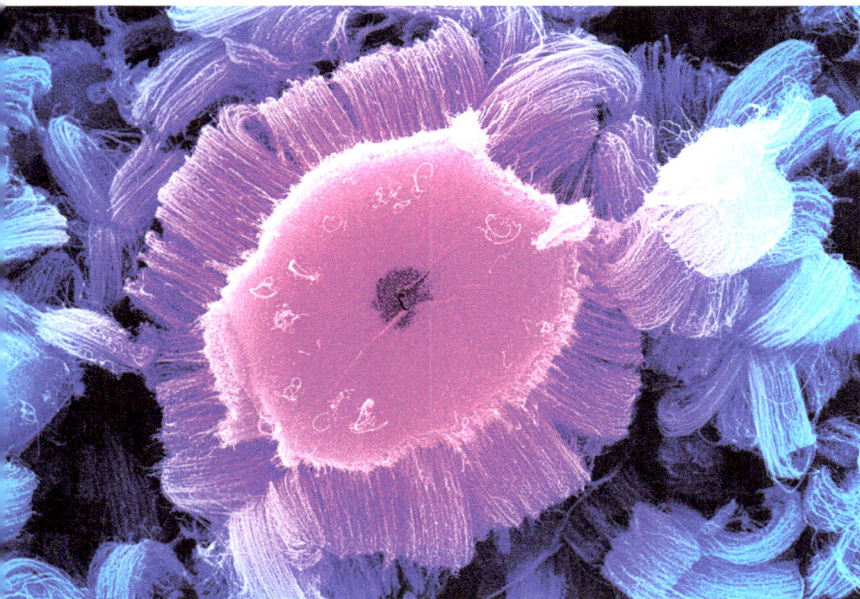

Medicine. Carbon nanotubes show great promise in medicine. Nanotubes are nontoxic, and they are not easily broken down by the digestive system. But, they are small enough to enter directly into living cells. Drug manufacturers can attach molecules to the outside of the carbon lattice or lock them inside it, to be released once the nanotube enters the cell. In this way, carbon nanotubes may enable more targeted medicine, with drugs delivered only to the cells that need them.

SPACE ELEVATOR

Astronauts, cargo, and space probes still travel to space much the same way they did 60 years ago—by rocket. Today's rockets are much improved from the ones used at the beginning of the space age. But rocket launches will always be expensive, loud, dependent on the weather, and somewhat dangerous. Rocket payloads must be somewhat compact to maintain the aerodynamic shape of the rocket. They must be below a certain weight, so the rocket can carry enough fuel to bring them to their destination. Payloads must also be resistant to the extreme forces felt during launch.

Science-fiction writers have long dreamt of a shortcut: the space elevator. A space elevator is a thick cable or tower that would extend from Earth's surface all the way into space. Special cars would transport people, cargo, fuel, parts, and products up and down the elevator. Steel cable is far too heavy and weak for such a mind-bending construction project. But strong, lightweight carbon nanotubes could one day make the space elevator possible.

People have come up with different ideas about how to build a space elevator. One idea involves moving an **asteroid** into *geostationary* orbit around Earth. Some 22,000 miles (36,000 kilometers) above the equator, the speed of an orbiting object matches that of Earth's rotation—causing the object to essentially hover over one spot on the ground. An asteroid in such an orbit could serve as a counterweight to support the cable, a source of raw materials, a base for the elevator's construction, and a space station once the elevator is complete.

To build such a space elevator, a robotic base would manufacture nanotubes from carbon mined on the geostationary asteroid. The base would construct a nanotube cable and project it through space toward Earth. Once the cable reached Earth, engineers would anchor it to the surface. Robots would climb up and down this initial cable, braiding more nanotube cables around it.

With a space elevator, spacecraft could be fueled, assembled, and even manufactured in space. Free from the punishing forces of launch, spacecraft could be larger and more delicate. They would no longer require huge amounts of fuel to escape Earth's **atmosphere**.

4 CONCRETE

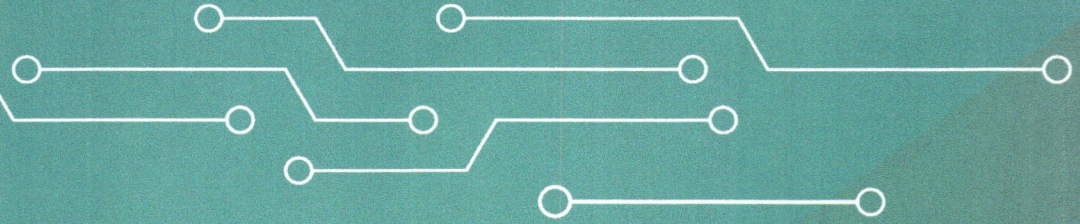

BUILDING THE PAST... AND THE FUTURE

Wherever you are, concrete is nearby. The foundation of the building you're in, the sidewalk, and the footing of the nearby fence are all made of concrete. The modern world is built on concrete. Over 170 tons (150 metric tons) of new concrete are produced every minute.

You may not think of concrete when you think of high-tech materials, but concrete revolutionized construction when it was developed by the ancient Romans thousands of years ago. The Romans built an empire on the stuff, creating some structures that survive today.

Modern advances in concrete are no less revolutionary. Scientists are working to make concrete stronger, longer-lasting, and more environmentally friendly.

CONCRETE HISTORY

The ancient Romans developed cement and concrete and used them extensively in building. To make cement, the Romans mixed slaked lime (lime to which water has been added) with a volcanic ash called *pozzuolana.* The result was so durable that several of their bridges, buildings, and aqueducts still stand some 2,000 years later.

Mortar lives on.
People lost the art of making cement after the fall of the Roman Empire in the A.D. 400's. But, builders continued to use simple mortars to bind brick and stone. These mortars consisted of slaked lime, sand, and a binder such as animal glue.

Rediscovery of cement. In 1756, John Smeaton, a British engineer, was tasked with rebuilding the Eddystone Lighthouse, near the south coast of Great Britain. He designed a tower of dovetailed blocks to withstand the harsh conditions. But the salty water and pounding waves would have quickly destroyed traditional mortar. Smeaton developed his own mixture containing *hydraulic lime,* a kind of powdered rock that sets when water is added. Smeaton's lighthouse outlasted the rocky outcrop on which it was perched. The tower was taken down in the 1880's, when engineers learned the rock beneath it was eroding away.

THE CLINKER REVOLUTION.

Smeaton's work set off a sensation in Great Britain, as inventors searched for recipes that were even stronger and easier to produce. For decades, manufacturers considered the *clinker* (hard pieces) produced in the making of cement to be waste. In 1843, the British businessman William Aspdin began to market a cement mixed with ground-up clinker. This clinkered cement produced concrete far stronger than any other cement on the market.

A concrete failure.
Aspdin was at best a poor businessman. He had a falling out with his father, Joseph, in whose factory he had discovered clinkered

cement. His invention was bound to be duplicated by competitors. But, he never filed for a patent, likely because he was still pretending to be part of his father's business. Clinkered *portland cement* exploded in popularity as a building material, but its inventor was left behind.

ENGINEERING CHALLENGE: DIFFICULT AND DIRTY

Concrete is everywhere, and its uses seem limitless. But the use of concrete also involves serious problems. First, concrete can be difficult to work with. It requires a precise mixture of cement, sand, gravel, and water, along with the right conditions to cure (harden) properly. If the ingredients or conditions are off, concrete will not last as long.

Even worse, concrete production can be dirty. To make portland cement, huge machines grind up rocks. The resulting powder is heated to a high temperature. Oil, gas, or powdered coal is used to produce the extremely high heat needed. All of these are fossil fuels that release carbon dioxide into the atmosphere when burned. The ground-up rock releases further carbon dioxide as it is heated. Eight percent of global greenhouse gas emissions come from the production of cement.

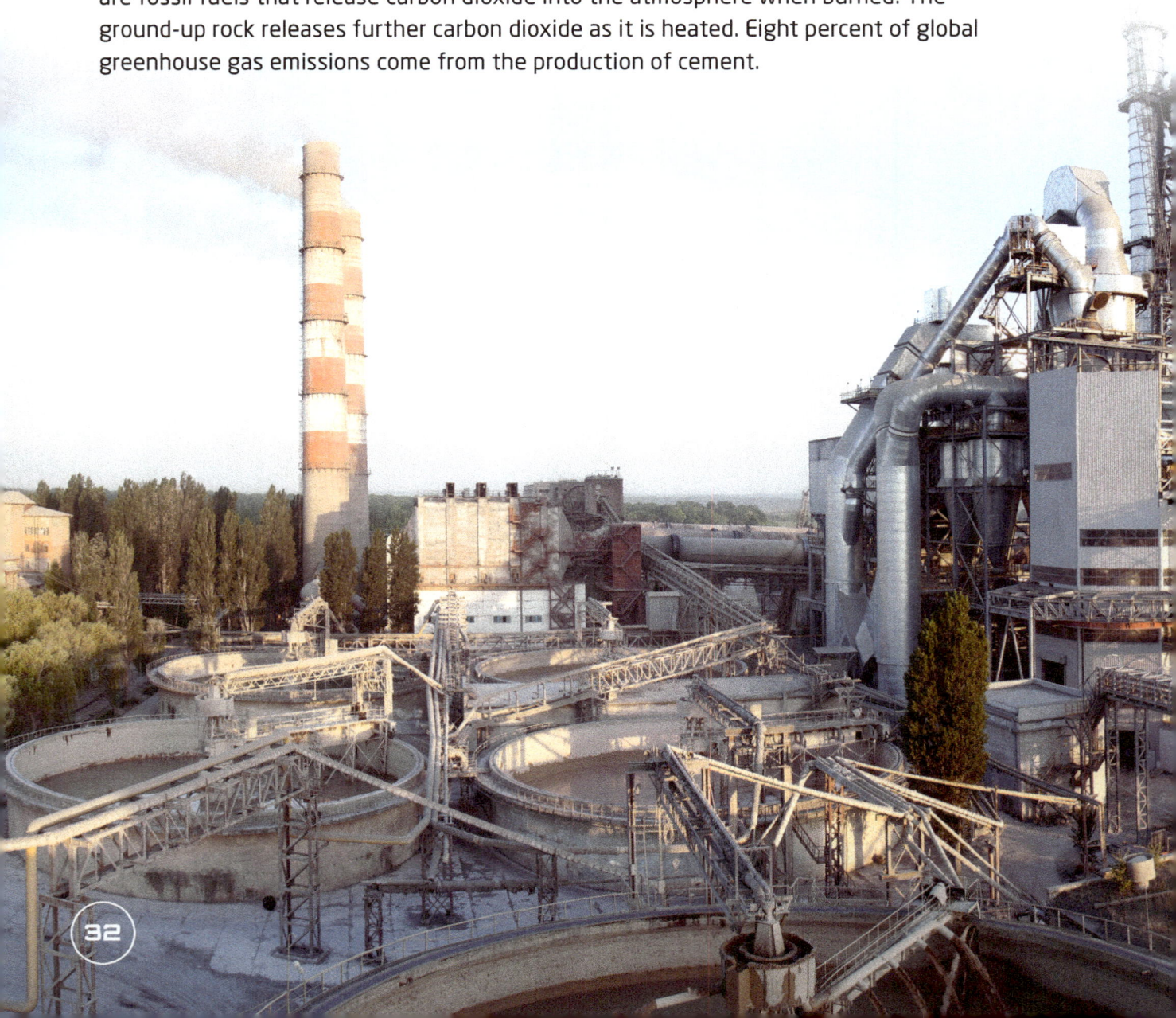

Cracks have also appeared in engineers' fondness for reinforced concrete, concrete cast around steel rods called rebar. Rebar is difficult to transport and work with. Care is required to set reinforced concrete correctly. The rebar must be placed to allow enough concrete to cover it. Rebar pieces must be secured tightly to prevent slipping when the concrete is poured. If the rebar winds up too near the surface of the concrete, moisture can seep in and corrode the steel. Steel expands as it corrodes, causing the surrounding concrete to crack. These cracks allow more moisture to reach the rebar, leading to further corrosion and cracking and eventually destroying the structure. Making the steel for rebar also consumes much energy, resulting in even more carbon dioxide emissions.

These disadvantages have led **architects** and engineers to favor alternate materials in some cases. But concrete will remain an important building material. For this reason, materials scientists are developing ways to make concrete stronger, more durable, and more environmentally friendly.

CONCRETE DREAMS

Salty ocean water can punish modern concrete. But many ancient Roman concrete docks and piers are still standing after 2,000 years. Researchers led by Marie Jackson of the University of Utah found that seawater actually strengthened this concrete. Over the years, the salty water percolates into the concrete and dissolves small amounts of pozzuolana and other components. Exotic minerals made up of the dissolved materials, sea salts, and water form in these spaces. These minerals resist further breakdown by salt water, strengthening the concrete century after century. Researchers are developing concretes that will take advantage of this process, helping to build stronger piers and seawalls.

Bulking additives. Concrete contains sand and gravel, which are mined in open pits and trucked to the building site. But, builders can include additives to reduce the need for these materials. Such additives include slag and ash from power plants and steel mills, crushed glass, and sawdust. Much of this material would otherwise be discarded in landfills. And, some of it may be available nearer the construction site, reducing the emissions involved in transportation.

Fiber-reinforced concrete. Fibers made of steel, fiberglass, or other **synthetic** materials can be mixed into concrete before it is cast. The resulting fiber-reinforced concrete is stronger than unreinforced concrete and easier to work with than concrete reinforced with rebar. It is better for the environment in most cases, since its manufacture uses less or no steel.

CO$_2$ncrete. Hydraulic cement is the dominant binder in concrete production, but it is not the only option. Some companies are developing concretes that do away with cement entirely. Instead, they mix waste slag in with the other concrete materials and inject the mixture with carbon dioxide gas captured from fossil-fuel emissions. The carbon dioxide and slag bond to harden the concrete. And, the carbon dioxide is kept out of the atmosphere for at least the life of the structure.

Smog-eating concrete. Traditional concrete can get dingy and dirty over time. The concrete forms of the Jubilee Church in Rome contain a special compound that keeps them pristine white. After the church was built, it was discovered that the compound "eats" air pollution. It traps pollutants while sunlight splits them into harmless chemicals.

5 METAMATERIALS

UNEXPECTED PROPERTIES

The way a material interacts with its environment is largely governed by its chemical makeup. For example, metal is shiny because its atoms have free electrons that interact with incoming light, reflecting their own light. Likewise, a red *pigment* (coloring agent) appears red because its molecules absorb other colors of light and reflect red light.

But reflection and absorption aren't the only way to deal with light and other incoming waves. Specialized materials called *metamaterials* can scatter incoming waves or cause them to interact with one another, producing some highly unexpected properties. These properties have more to do with physics than chemistry—they result from the material's microscopic structure, rather than its chemical makeup.

Metamaterials can have unexpected colors or even appear invisible—and that's just for starters. They may help us take clearer photographs, but they could also lead to the creation of science fiction-style cloaking devices.

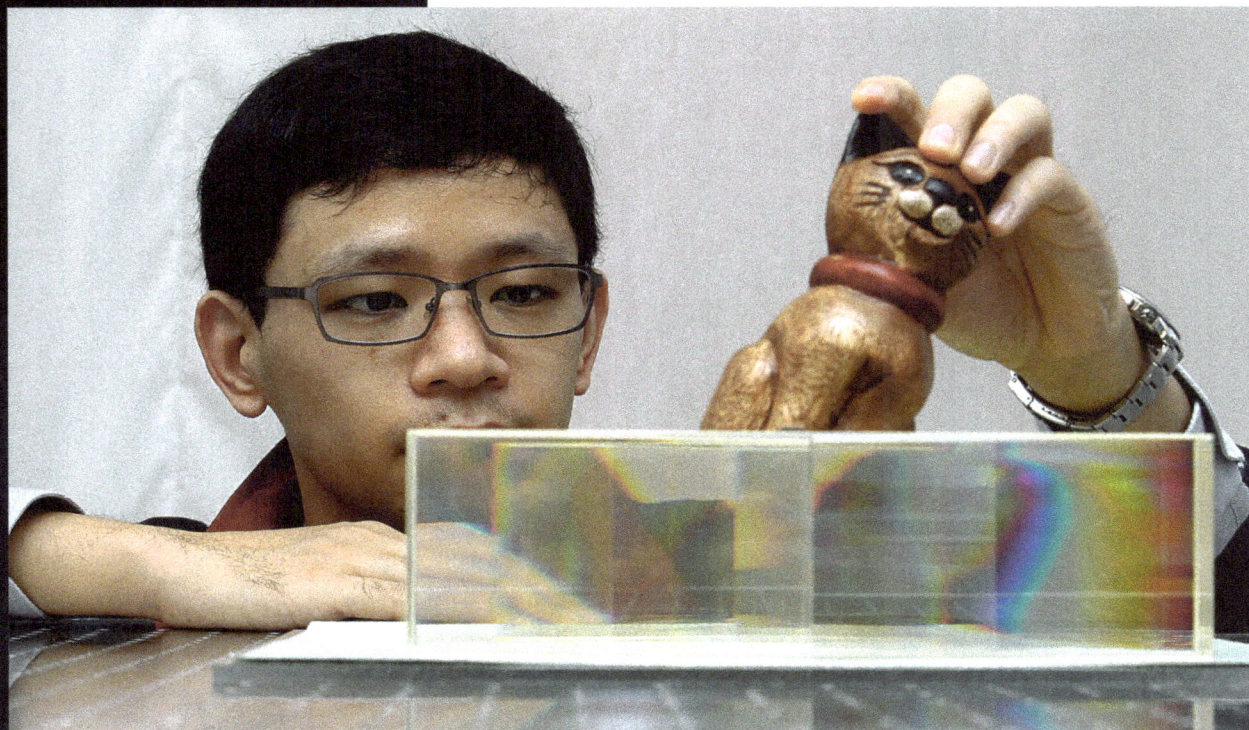

METAMATERIAL BASICS: WORKING WITH WAVES

Several kinds of energy travel in waves. Waves have special properties. When waves meet, they can combine to make larger waves, or they can cancel each other out. For example, noise-canceling headphones work by sensing incoming sound waves and producing sounds to cancel them before they reach the wearer's ears.

Interference

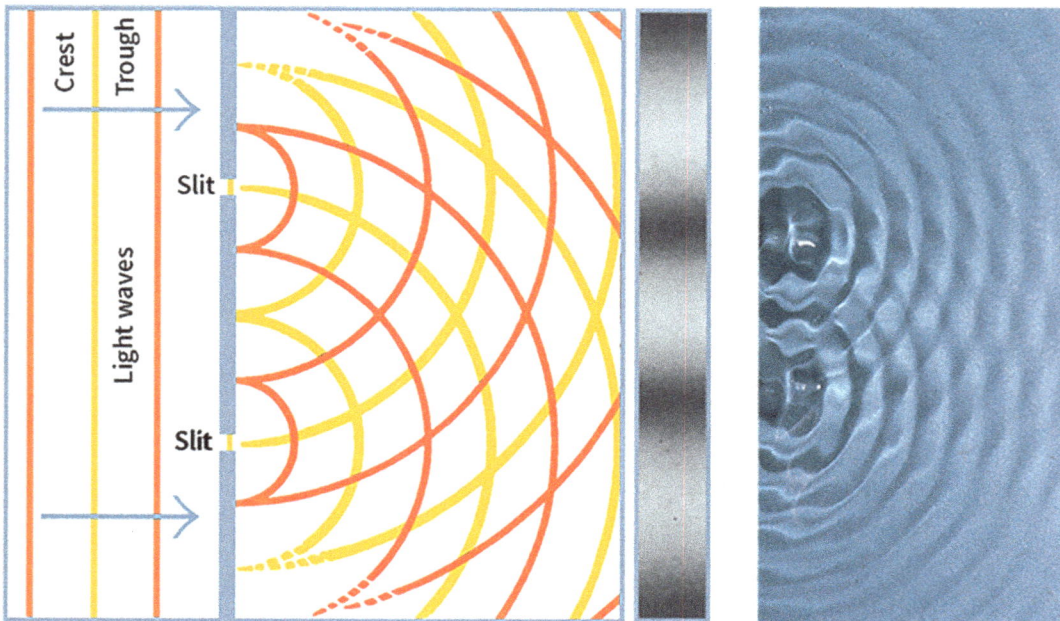

Crest

Trough

Light waves

Slit

Slit

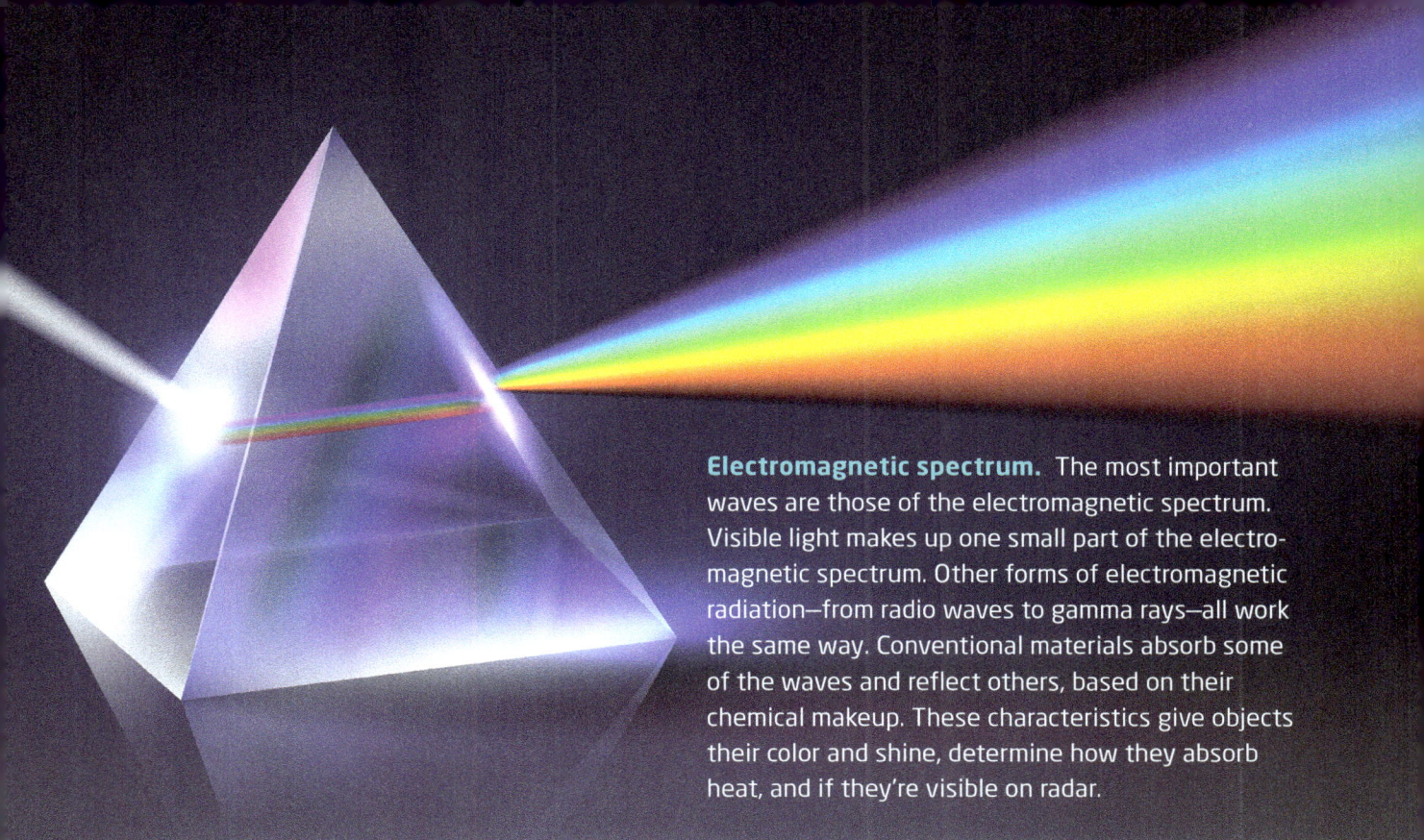

Electromagnetic spectrum. The most important waves are those of the electromagnetic spectrum. Visible light makes up one small part of the electromagnetic spectrum. Other forms of electromagnetic radiation—from radio waves to gamma rays—all work the same way. Conventional materials absorb some of the waves and reflect others, based on their chemical makeup. These characteristics give objects their color and shine, determine how they absorb heat, and if they're visible on radar.

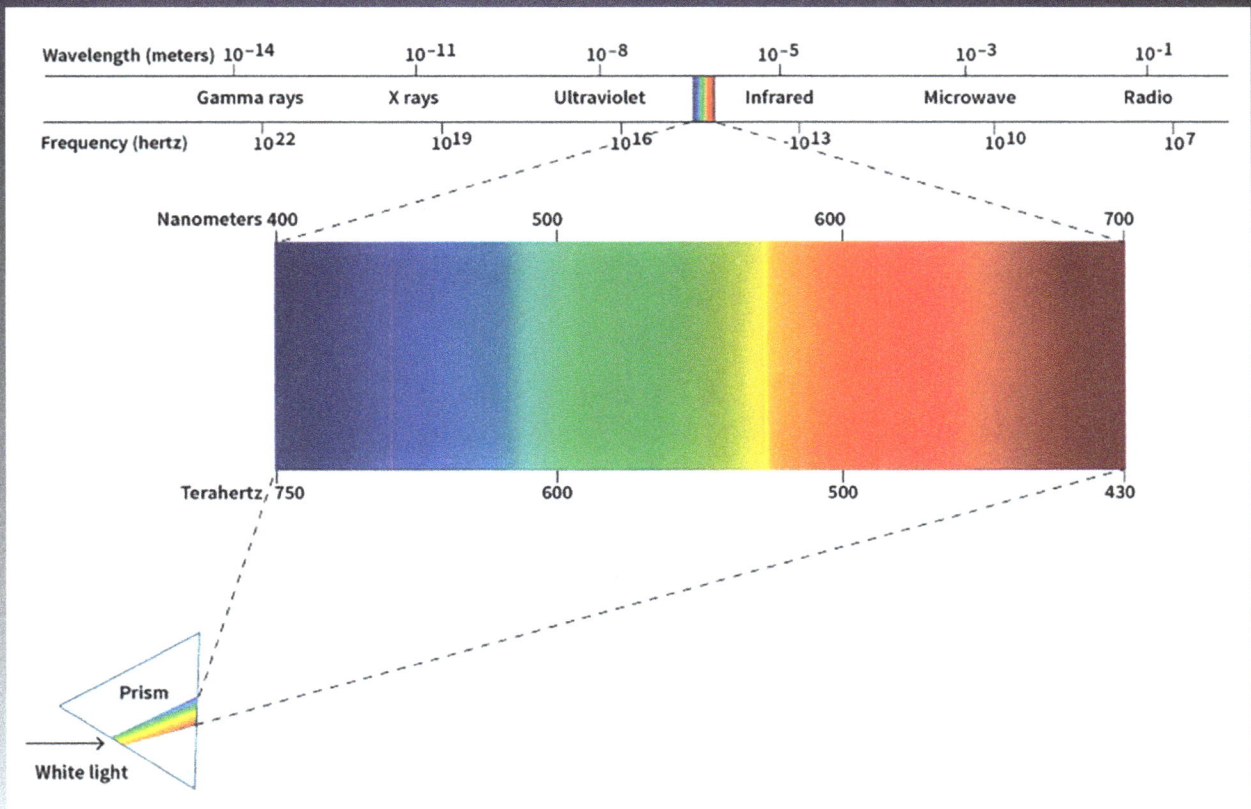

Wavelength (meters)	10^{-14}	10^{-11}	10^{-8}	10^{-5}	10^{-3}	10^{-1}
	Gamma rays	X rays	Ultraviolet	Infrared	Microwave	Radio
Frequency (hertz)	10^{22}	10^{19}	10^{16}	10^{13}	10^{10}	10^{7}

Nanometers 400	500	600	700

Terahertz 750	600	500	430

Prism

White light

Structural coloration. When is a colorless material not colorless? When it contains extremely tiny structures that scatter incoming light or cause it to interfere with itself. This effect is called structural coloration. The shape of the material, not its chemical makeup, creates the color.

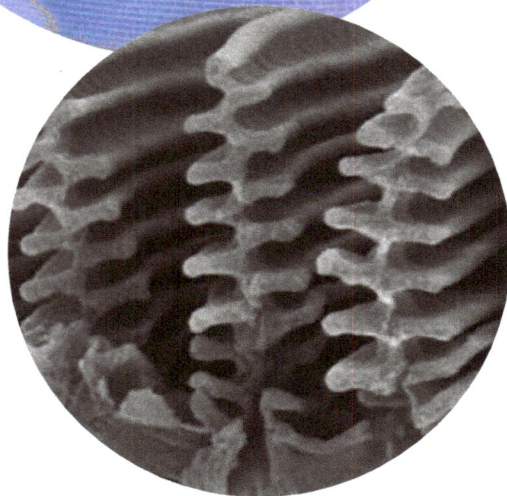

Metamaterials in nature? Materials scientists are just beginning to play with structural coloration in metamaterials. But, some living things exhibit structural coloration in their tissues naturally. This effect often produces vivid or shimmering blues and greens. The feathers of a peacock, the jewellike wings of some beetles, and even blue and green eyes in people are examples of structural coloration.

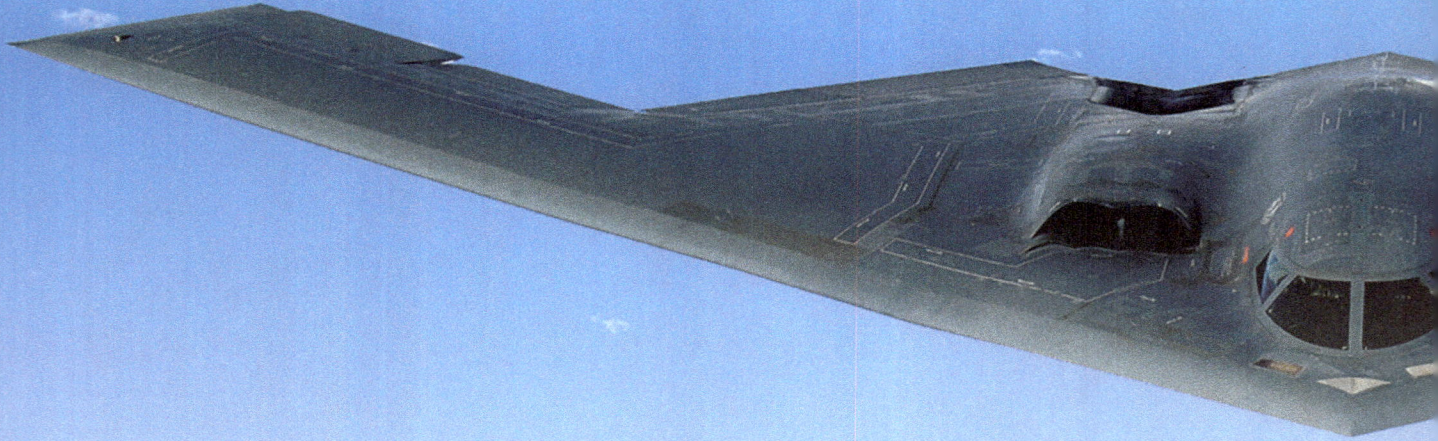

ENGINEERING CHALLENGE: THE LIMITS OF STEALTH

Metamaterials may have important applications in the field of stealth. Stealth aircraft can penetrate deep into enemy territory to strike without being detected. The major concern in stealth is evading radar technology. Such technology detects craft by bouncing radio waves off their surfaces.

Stealth airplanes come in a variety of shapes with strange curves and facets. These shapes are designed to deflect radio waves away, rather than reflect them to radar detectors. But, they come with significant design tradeoffs. Stealth airplanes are less maneuverable and stable than their unstealthy counterparts.

Odd shapes are not the only technology helping aircraft evade detection. The United States B-2 Spirit bomber, for example, makes use of a special radar-absorbing paint. However, the paint is extremely temperamental. It even washes off in the rain. B-2's must therefore be kept in climate-controlled, dust-filtered hangars. Technicians regularly remove and re-apply the stealth coating. But the paint is toxic, so the technicians must wear protective equipment. All of this

makes the planes extremely expensive to maintain, even when they are sitting in the hangar.

And after all that, advanced radar—and perhaps even basic radar—may be able to detect stealth aircraft in the right circumstances. The United States' F-117 Nighthawk was the first true stealth airplane. But, during the Persian Gulf War of 1991, British destroyers were able to detect the aircraft from 40 miles (60 kilometers) away using conventional radar. When this information leaked to the British press, it caused a minor diplomatic crisis between the country and the United States. In 1999, Serbian forces used espionage and multiple radar systems to track and shoot down an F-117.

Military strategists think that metamaterials could be the next breakthrough in stealth technology. Metamaterials might be used in better radar-resistant coatings for stealth aircraft. And metamaterials can manipulate more than just radio waves, offering the tantalizing promise of fully cloaked vehicles.

USES OF METAMATERIALS

Antireflection glass is a simple but exciting use of metamaterials technology. Glass can exhibit undesirable reflection and glare at wide angles. Plastic coatings can minimize this glare, but they reduce light penetration and can scratch. A company called Edgehog has created glass with built-in reflection protection using metamaterial technology. This glass is useful for everything from cameras to solar panels.

Invisibility cloaks. Metamaterials can be used to bend light around an object, rendering it invisible. A research team led by scientists from Duke University created a device that cloaked an object inside from microwaves. Cloaking an object from a broader range of the electromagnetic spectrum presents a much greater challenge. Metamaterials must be specially designed to handle a small range of wavelengths. A metamaterial that works on microwaves, for example, would not work with visible light.

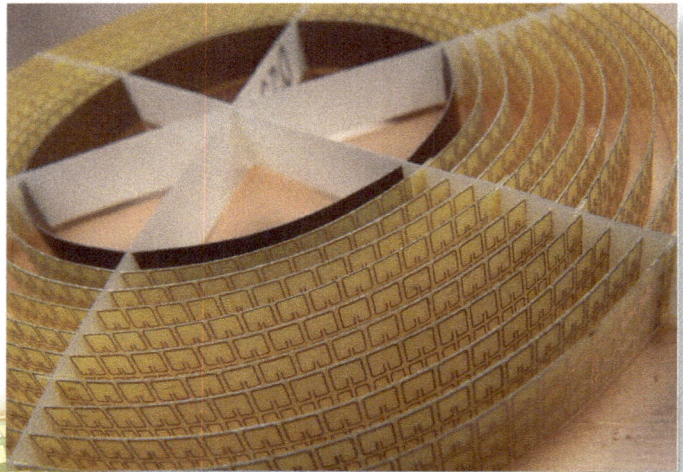

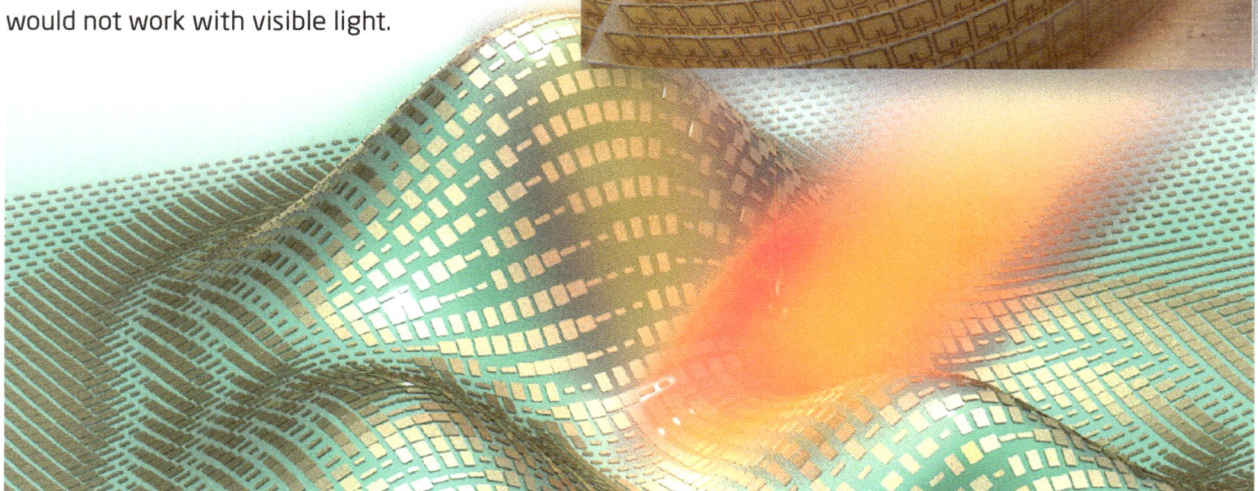

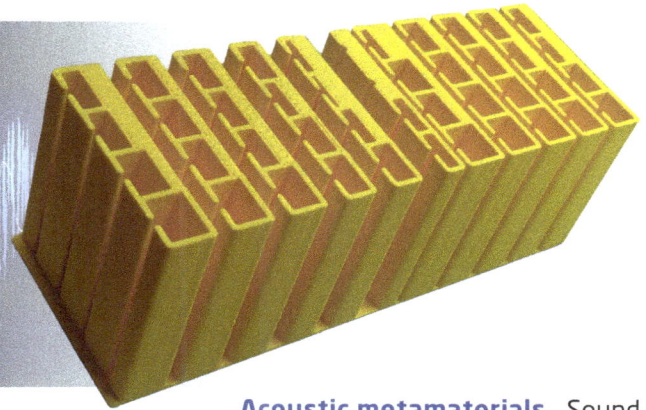

Acoustic metamaterials. Sound travels in waves, too. Acoustic metamaterials could be designed to amplify, focus, or even reduce sound waves. For example, researchers at the Hong Kong University of Science and Technology are studying a metamaterial acoustic screen. The screen has tiny holes that air can pass through but that discourage the passage of sound waves. Such screens could be used to reduce engine noise and other forms of noise pollution.

Safer cars. Engineers at the University of Glasgow, Scotland, have developed a **3D-printed** metamaterial using plastics and carbon nanotubes that is highly resistant to impacts The engineers hope the new metamaterial will be useful in making safe, lightweight cars.

Miniaturized antennas. Perhaps the most overlooked yet vital piece of our interconnected world is the lowly antenna. No longer prominently displayed atop cell phones and homes, modern antennas are flattened coils of wire. But as smartphones and other electronics have shrunk, so has the space available for antennas. Engineers are working with metamaterials to make smaller and stronger antennas.

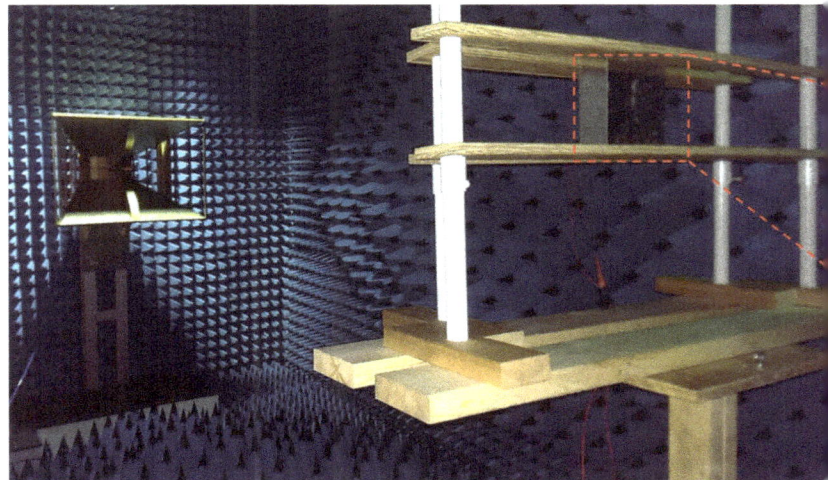

ENGAGE YOUR READER

Nonfiction writing often includes subject-specific vocabulary terms. Knowing the words related to the topic helps us understand the text itself.

When good readers come upon words they don't know well, they pause and try to figure them out. One tool they use is the glossary, like the one on page 4. Not every word can be defined in a glossary, though!

Authors know this, so they leave clues about words in the text. Next time you encounter a challenging word, stop and look for information about its meaning in the surrounding sentences. Sometimes authors define the term right there in the text! Other times, they'll compare the term to something you may already know. Authors even use punctuation like commas or dashes to clue you in to a word's meaning.

INSTRUCTIONS

1. Consider the list of challenge words and identify where each is used in the text. You can use the Index on page 48 to help you locate each term.

2. Explain how the author described each word. Ask yourself "what is happening in the text?" or "how is this word being used?" as you search for clues about their meanings.

3. Create your own definitions of the words. Don't just copy the dictionary definitions. Instead think about how you would tell a friend what each term means.

4. Add a visual representation for each word. Think about what you could draw that will help you remember what the words mean.

CHALLENGE WORDS

- Aerogel
- Pores
- Insulator
- Carbon

- Graphene
- Carbon nanotube
- Concrete
- Metamaterial

EXAMPLE

Challenge Word	Page(s)	Author's Description	Personal Definition	Visual Representation
Aerogel	7-11	- an artificial, dry solid known for its exceptionally low density - "frozen smoke" and "frosted glass" - many pores	A material that is incredibly lightweight because it is mainly made of air. It can be turned into many useful products.	
Pores				

INDEX

www.ingramcontent.com/pod-product-compliance
Lightning Source LLC
Chambersburg PA
CBHW061419090426
42744CB00018B/2074